The Ultimate Guide to Eclipses

Understanding, Observing, and Appreciating Celestial Phenomena

Daniel Mario

Copyright

All rights reserved. No part of this publication may reproduced, distributed or transmitted in any form or by any means including photocopying, recording or other electronic or mechanical method, without the prior written permission of the publisher, except in the case of brief quotation embodied in critical reviews and certain other non commercial uses permitted by copyright law.

Copyright © Daniel Mario,2024.

This page belongs to

Table of Contents

Introduction
Chapter 1
The Basics of Eclipses
    What is an Eclipse?
        Solar Eclipses
        Lunar Eclipses
    Historical Significance
    Cultural Interpretations
    Ancient Observations
    Celestial Mechanics
    Orbital Dynamics
    Eclipse Prediction
Chapter 2
Types of Eclipses
    Solar Eclipses
        Total Solar Eclipses
        Partial Solar Eclipses
        Annular Solar Eclipses
    Lunar Eclipses
        Total Lunar Eclipses
        Partial Lunar Eclipses
        Penumbral Lunar Eclipses
Chapter 3
The Science Behind Eclipse
    Celestial Mechanics
        Orbital Dynamics
        Eclipse Prediction

    Phenomena Explained
        Umbra and Penumbra
        Eclipse Cycles and Frequencies
Chapter 4
Observing Eclipse Safely
    Solar Eclipse Safety
        Protective Eyewear
        Viewing Techniques
    Lunar Eclipse Viewing
        Naked-Eye Observation
        Binocular and Telescope Use
    Beyond Eclipses: Related Celestial Events
        Transits and Occultations
        Meteor Showers
Chapter 5
Beyond Eclipses: Related Celestial Events
    Transits and Occultations
        Venus Transits
        Star Occultations
    Meteor Showers
        Annual Meteor Showers
        Viewing Tips and Best Practices
Conclusion

Introduction

Welcome to "The Ultimate Guide to Eclipses: Understanding, Observing, and Appreciating Celestial Phenomena." In this comprehensive ebook, we embark on a journey through the captivating world of eclipses, exploring their significance, science, and splendor.

Unveiling the Mysteries of the Cosmos

Eclipses have always captured humanity's imagination, inspiring awe and amazement as the sun, moon, and Earth align in breathtaking displays of cosmic ballet. Eclipses have been viewed, researched, and revered by people throughout history, from ancient civilizations to current scientists, for their profound symbolism and scientific significance.

A Beacon of Knowledge and Understanding

Our objective is to uncover the secrets of eclipses, putting light on their underlying principles and complex dynamics. Drawing inspiration from the writings of renowned astronomer Neil deGrasse Tyson, we attempt to communicate complicated scientific concepts in an approachable and interesting format.

Navigating the Cosmos with Confidence

Whether you are a beginner or an experienced astronomer, "The Ultimate Guide to Eclipses" will provide you with the knowledge and skills you need to safely and responsibly watch and appreciate these celestial occurrences. We simplify the physics of eclipses with clear, short explanations and visual examples, allowing you to confidently explore the sky.

Setting the Stage for Exploration

As we embark on this celestial trip together, we welcome you to experience the wonder and beauty of eclipses, marveling at the cosmic dance that unfolds in front of us. With each turn of the page, may you develop a better comprehension of the universe and a greater respect for the wonders that exist beyond our planet.

Embracing the Wonder of the Cosmos

Join us on an expedition through space and time, discovering the enduring appeal of eclipses and revealing the mysteries of the universe. Allow us to embark on a journey of discovery, where the wonders of the universe await those who dare to imagine and explore.

In the pages that follow, prepare to embark on an exhilarating adventure through the celestial wonders of eclipses. With the spirit of curiosity as our guide and the wisdom of the cosmos as our compass, let us journey together into the heart of the unknown, where the mysteries of the universe await our discovery.

Chapter 1

The Basics of Eclipses

In this foundational chapter, we will look at the underlying principles of eclipses and explore the mysteries of these celestial phenomena that have captivated humanity's imagination for millennia. To comprehend the essence of these celestial spectacles, we shall examine the mechanics of solar and lunar eclipses in addition to their historical significance and cultural beliefs.

What is an Eclipse?

When three celestial bodies—the sun, moon, and Earth—align in space and cast shadows on one another, an eclipse occurs. There are two main kinds of eclipses: solar and lunar, which result from this complex dance of light and shadow.

Solar Eclipses

A solar eclipse occurs when the moon moves in front of the sun, illuminating part of the planet's surface with shadows. This astronomical alignment causes a momentary obscuration of the sun, turning the day into an eerie twilight as the moon partially or blocks the sun's disk.

Lunar Eclipses

A lunar eclipse happens when the Earth passes between the sun and the moon, leaving a shadow on the lunar surface. The moon is enveloped in an ethereal glow during this fascinating occurrence, which passes through the shadow of Earth and experiences a series of astounding changes.

Historical Significance

Eclipses have been deeply significant to societies all across the world throughout history, evoking equal parts awe, fear, and veneration. Eclipses have been viewed as omens, portents, and heavenly messages from the gods by communities spanning from ancient times to contemporary times.

Cultural Interpretations

Eclipses were considered in ancient Mesopotamia to be signs of the approaching end of empires and the gods' displeasure. Similar to this, eclipses were connected in ancient China with the sun or moon being devoured by a celestial dragon, which led to rites and sacrifices to appease the cosmic powers involved.

Ancient Observations

Ancient civilizations carefully documented and monitored eclipses, noting their predicted patterns and cyclical nature, despite their limited scientific understanding. The foundation for the methodical study of eclipses and their celestial mechanics was built by early skywatchers, from the Babylonian astronomers of Mesopotamia to the Maya priests of Mesoamerica.

Celestial Mechanics

Celestial mechanics, the study of how the laws of physics control the motions of celestial bodies and determine when and how often eclipses occur, is necessary to fully understand the complexities of eclipses.

Orbital Dynamics

Orbital dynamics is a fundamental concept in eclipse studies, as it describes how the sun, moon, and Earth move through their elliptical orbits in space. Astronomers can make very accurate predictions about the frequency and features of eclipses by examining the orbital parameters and relative positions of various celestial entities.

Eclipse Prediction

Astronomers have created complex models and algorithms to forecast the occurrence of eclipses years, decades, or even centuries in the future through centuries of observation and mathematical research. Through the use of these predictive methods, astronomers can predict with previously unheard-of precision the exact timing, duration, and geographic visibility of forthcoming eclipses of the sun, moon, and Earth.

We have covered a lot of ground in this insightful chapter, including the basic ideas and historical significance of eclipses. From the intricate mechanics of solar and lunar eclipses to the cultural interpretations and ancient observations that have affected our understanding, we have set out on a journey to explore the mysteries of the universe. Join us as we continue our journey across the cosmos, where eclipses await those who dare to seek knowledge and appreciate the night sky.

Chapter 2

Types of Eclipses

Building on the knowledge gained in first Chapter, we now shift our attention to the various eclipses that grace the cosmic stage. From total solar eclipses to partial lunar eclipses, we look at the distinct qualities and compelling beauty of each form of eclipse, shedding light on the complexities of their celestial dance.

Solar Eclipses

Total Solar Eclipses

Total solar eclipses are among the most spectacular celestial phenomena, occurring when the moon obscures the sun, casting a shadow called the umbra on the Earth's surface. As the moon perfectly aligns with the sun, the sky darkens, revealing the sun's corona in a stunning show of solar splendor. Total solar eclipses are unusual and transitory events that provide observers with a momentary glimpse into the solar system's cosmic grandeur.

Partial Solar Eclipses

Partial solar eclipses occur when the moon obscures only a piece of the sun's disk, resulting in a beautiful celestial tableau in which the sun looks to be bitten by a cosmic crescent. While not as dramatic as total solar eclipses, partial eclipses continue to captivate onlookers with their ominous atmosphere and celestial spectacle, acting as a reminder of the cosmic forces at work in our solar system.

Annular Solar Eclipses

During annular solar eclipses, the moon appears smaller than the sun, creating a ring of sunlight known as the "ring of fire" around the moon's dark silhouette. This fascinating phenomenon occurs when the moon reaches apogee, its farthest point from Earth, and appears slightly smaller in the sky. Annular solar eclipses provide an unparalleled opportunity to observe the exquisite interplay of light and shadow in the cosmic dance of celestial bodies.

Lunar Eclipses

Total Lunar Eclipses

Total lunar eclipses happen when the Earth passes squarely between the sun and the moon, leaving a shadow on the lunar surface. As the moon moves through the Earth's shadow, it turns a dramatic scarlet color, gaining the nickname "blood moon." Total lunar eclipses capture onlookers with their uncanny beauty and celestial drama, providing a glimpse into the cosmic ballet that plays out in the night sky.

Partial Lunar Eclipses

Partial lunar eclipses occur when only a section of the moon passes through the Earth's shadow, darkening the lunar surface. While not as dramatic as total eclipses, partial lunar eclipses provide viewers with an enthralling celestial sight as the moon becomes delicate shades of gray and crimson against the night sky.

Penumbral Lunar Eclipses

Penumbral lunar eclipses occur when the moon passes through the penumbra, which is the outermost portion of the Earth's shadow. During these eclipses, the darkening of the lunar surface is slight and often difficult to detect, resulting in a less dramatic celestial show. Penumbral lunar eclipses remind us of the exquisite interplay of light and shadow in the cosmic dance of the sun, moon, and Earth.

In this chapter, we have looked at the various types of eclipses that grace the cosmic stage, from the breathtaking majesty of complete solar eclipses to the delicate elegance of penumbral lunar eclipses. By delving into the distinct characteristics and celestial mechanics of each type of eclipse, we have gained a greater appreciation for the glories of the cosmos and the delicate dance of celestial bodies that takes place in the night sky. Join us as we continue our adventure through the cosmos, where eclipse secrets await those who dare to explore the universe's depths and revel in the beauty of celestial ballet.

Chapter 3

The Science Behind Eclipse

In this chapter, we go on a scientific journey through eclipses, diving into the celestial mechanics and physical laws that control these fascinating phenomena. From the orbital dynamics of the Earth, moon, and sun to the intricate interplay of light and shadow, we uncover eclipse riddles and gain a better grasp of their cosmic significance.

Celestial Mechanics

Orbital Dynamics

Eclipses are defined by the complicated dance of celestial bodies as they travel through space in elliptical orbits. The Earth orbits the sun, and the moon orbits the Earth; their relative positions affect the occurrence and characteristics of eclipses. Astronomers can accurately anticipate the date and occurrence of eclipses by examining their orbital properties and trajectories.

Eclipse Prediction

Astronomers have created complex models and algorithms for predicting eclipses over centuries of observation and mathematical study. By calculating the precise alignments of the sun, moon, and Earth, these predictive techniques allow astronomers to forecast the date, duration, and geographic visibility of eclipses years, decades, and even centuries in advance. This predictive power enables astronomers to plan and conduct eclipse observations with precision and accuracy.

Phenomena Explained

Umbra and Penumbra

The umbra and penumbra, two regions of shadow cast by celestial bodies, are crucial to comprehending eclipses. The umbra is the core, the darkest area of the shadow where the light source is fully hidden, causing total eclipses. The penumbra surrounds the umbra and is a lighter, outer region where just a portion of the light source is blocked, causing partial eclipses. Understanding the dynamics of these shadow zones is critical for forecasting and viewing eclipses.

Eclipse Cycles and Frequencies

Eclipses follow predictable cycles and frequency due to the intricate interactions of the Earth, moon, and sun. The Saros cycle, for example, is approximately 18 years and 11 days during which a succession of eclipses occur in a predetermined order. Astronomers may predict eclipses years, decades, and even millennia in advance by studying these cycles and frequencies, yielding vital insights into the dynamics of the solar system and the evolution of celestial phenomena.

We have examined the laws of science and celestial mechanics governing eclipses in this insightful chapter, illuminating the complex interactions between light and shadow in the solar, lunar, and Earthly dance. We have taken a huge step in unlocking the secrets of the universe and embracing the beauty of the cosmic ballet by unraveling the mysteries of eclipses and gaining a better grasp of their underlying mechanics. Join us as we continue our adventure into the cosmos, where eclipses await those who dare to explore the depths of the cosmos and appreciate the magnificence of the night sky.

Chapter 4

Observing Eclipse Safely

In this chapter, we will look at how to safely and responsibly observe eclipses. From solar eclipse safety precautions to lunar eclipse viewing techniques, we offer vital advice to guarantee that watchers can enjoy the glories of eclipses without jeopardizing their sight or safety. By emphasizing the significance of proper equipment, procedures, and preparation, we enable enthusiasts of all skill levels to view these celestial events in all their majesty.

Solar Eclipse Safety

Protective Eyewear

One of the most important components of watching a solar eclipse is to protect your eyes from the sun's powerful radiation. Direct exposure to the sun without sufficient eye protection can result in irreversible eye damage, such as solar retinopathy. To protect your vision, use approved solar viewing glasses or sun filters that fulfill the ISO 12312-2 safety standards. These specialist glasses and filters prevent dangerous UV, visible, and infrared light, allowing you to safely view the sun during an eclipse.

Viewing Techniques

When witnessing a solar eclipse, it is critical to use protective glasses and adopt suitable viewing procedures. Instead of staring directly at the sun, which can strain your eyes and cause damage, employ indirect viewing devices like pinhole projectors or solar viewing telescopes. These gadgets allow you to safely watch the sun's image projected onto a surface, such as a piece of cardboard or a white sheet of paper, without subjecting your eyes to hazardous radiation.

Lunar Eclipse Viewing

Naked-Eye Observation

Unlike solar eclipses, lunar eclipses are safe to watch with the naked eye. A total lunar eclipse is a spectacular celestial show that is visible to viewers all around the world. The moon takes on a dramatic reddish tint as it passes into the Earth's shadow. Simply choose a comfortable spot with a clear view of the night sky and marvel at the stunning beauty of the lunar eclipse as it unfolds in front of your eyes.

Binocular and Telescope Use

Consider utilizing binoculars or a telescope to improve your viewing experience if you want to see the lunar eclipse up close. By using binoculars, you can see the moon's surface more clearly and see craters, mountains, and other lunar features. On the other hand, telescopes provide enlarged views of the moon's surface and enable more in-depth study of its complex terrain and geography. Just make sure to steady your telescope or binoculars with a tripod or sturdy mount to reduce image shake while you observe.

Beyond Eclipses: Related Celestial Events

Transits and Occultations

The universe has a plethora of other celestial events to investigate and witness in addition to eclipses. A celestial body transits when it passes in front of another. For example, Venus transits the solar disk in pairs every eight years, and then it does not happen for more than a century. On the other hand, occultations happen when one celestial body is blocked from view by another, like when a planet passes in front of a far-off star or a star vanishes behind the moon. These occasions provide exceptional chances to see uncommon and amazing celestial occurrences and broaden our knowledge of the universe.

Meteor Showers

Another amazing astronomical event that takes place when Earth travels through an asteroid or comet's debris trail is a meteor shower. Meteors, often called shooting stars, are flashes of light formed when debris particles melt and hit with Earth's atmosphere. Numerous meteor showers occur in the night sky every year, each having unique radiant points and peak viewing hours. Astronomers can learn more about the origins and development of the solar system by studying the makeup and behavior of comets and asteroids through the observation of meteor showers.

We have discussed the significance of properly and responsibly witnessing eclipses in this important chapter, offering pointers and advice to amateurs and experts alike. We have talked about how important it is to use the right equipment and protect your eyes when viewing celestial events, from solar eclipse safety measures to lunar eclipse viewing strategies. By adhering to these rules, you can enjoy a safe and pleasant cosmic adventure while witnessing the glories of eclipses and other astronomical occurrences without jeopardizing your vision. Join us as we continue our journey into the night sky, where the wonders of the cosmos await those who dare to look up and appreciate the beauty of the cosmic ballet.

Chapter 5

Beyond Eclipses: Related Celestial Events

As we come to the end of our voyage through the world of eclipses, we turn our attention to the multitude of linked celestial occurrences that illuminate the night sky. These fascinating occurrences, which range from planetary transits to meteor showers, provide astronomers and stargazers with equal opportunity to observe the wonders of the cosmos and expand their knowledge of the universe. We examine the variety of astronomical occurrences that extend beyond eclipses in this chapter, providing insights into their scientific value, observational methods, and significance.

Transits and Occultations

Venus Transits

The transit of Venus across the face of the sun is one of the rarest and most awaited celestial phenomena. Each pair of these transits is separated by more than a century, and they happen in pairs every eight years. A Venus transit occurs when the planet Venus appears as a tiny black dot on the solar disk and travels directly between the Earth and the sun. Astronomers may learn a great deal about Venus's composition and atmosphere from these transits, and they can also improve methods for finding exoplanets circling far-off stars.

Star Occultations

When one heavenly body moves in front of another, momentarily blocking out the other's light, this is known as an occultation. One of the most frequent kinds of occultation is a stellar occultation, in which a planet or asteroid briefly obscures a far-off star by passing in front of it. Astronomers can determine the size, shape, and makeup of celestial bodies as well as the dynamics and composition of their atmospheres by studying these occultations.

Meteor Showers

Annual Meteor Showers

Every year, as the Earth passes through the debris trail left by an asteroid or comet, meteor showers take place in the sky. The debris particles evaporate as they collide with the Earth's atmosphere, producing meteors, also known as shooting stars, which are flashes of light. Numerous meteor showers, each with its peak viewing hours and radiant spots, can be seen in the night sky every year. The Perseids in August, the Geminids in

December, and the Leonids in November are a few of the most well-known meteor showers.

Viewing Tips and Best Practices

To witness meteor showers, go to a dark place away from city lights with a clear view of the night sky. Bring a comfy chair or blanket to sit or lie down on, as meteor watching may be a relaxing activity that takes patience and perseverance. Allow your eyes to adjust to the darkness, then scan the sky for shooting stars, paying special attention to the shower's radiant point, where meteors appear to originate. Prepare to spend several hours observing meteor showers, which can create occasional bursts of activity punctuated with periods of silence.

In this captivating chapter, we have explored the diverse array of celestial events beyond eclipses, from planetary transits to meteor showers. We developed a greater appreciation for the wonders of the cosmos and the beauty of the night sky by digging into their significance, observing techniques, and scientific importance. As we conclude our tour through the celestial heavens, we welcome you to continue exploring the universe's mysteries and basking in the majesty of the cosmic ballet that plays out above us. Join us as we look up, into the boundless expanse of space, where the wonders of the cosmos await those who dare to dream and explore.

Conclusion

As we come to the end of "The Ultimate Guide to Eclipses: Understanding, Observing, and Appreciating Celestial Phenomena," we reflect on our incredible voyage through the universe. From the fundamentals of eclipses to the physics underlying these celestial phenomena, we have delved into the mysteries of the cosmos and obtained a better grasp of the wonders that exist beyond our terrestrial domain.

Throughout our journey, we have marveled at the exquisite dance of the sun, moon, and Earth as they align in breathtaking displays of light and shadow. We have learnt about the various forms of eclipses, including total solar eclipses and penumbral lunar eclipses, as well as the celestial mechanics that determine their occurrence and features. We have also looked into the historical significance of eclipses, from ancient interpretations to modern scientific breakthroughs, and developed a new understanding of their cultural and cosmic value.

In addition to eclipses, we have investigated other astronomical phenomena such as planetary transits and meteor showers. These enthralling displays allow astronomers and stargazers to observe the majesty and grandeur of the universe, enhancing our connection to it and evoking a sense of wonder and awe.

As we come to the end of our voyage, we are reminded of the need to properly and responsibly witness eclipses, preserve our vision, and use suitable equipment and approaches. Whether we are watching a solar eclipse with protective eyewear or admiring a meteor shower under a starry night sky, we must always respect the celestial realm's strength and majesty.

But our adventure does not stop here. The cosmos invites us onward, inviting us to continue studying the universe's mysteries and unlocking the secrets of the night sky. With each observation and discovery, we get a little closer to solving the mystery of our cosmic origins and comprehending our place in the vastness of space and time.

We leave "The Ultimate Guide to Eclipses" with a renewed sense of amazement and inquiry, fuelled by the cosmos' unlimited beauty and complexity. Let us continue our adventure with open minds and hearts, embracing the boundless possibilities that lie beyond the horizon and aiming for the stars with unwavering enthusiasm and resolve.

Thank you for joining us on this exciting journey into the magnificent realms. May your nights be filled with wonder, and may the light of the stars lead your dreams.

Until we meet again beneath the canopy of the night sky, remember to look up, because the wonders of the cosmos await those who dare to dream and explore.

www.ingramcontent.com/pod-product-compliance
Lightning Source LLC
Chambersburg PA
CBHW070957220526
45471CB00007B/3076